A New Earth

The Core Cause of Climate Change

(And What You Can Really Do About it)

Other Books by Bill Wylson

Hieroglyphs, Golden Plates & Typos
Give Place in Your Heart
Three Minutes Eighteen Seconds
Elder Hammond and The Inspector
The Manger on the Mantle

A New Earth

The Core Cause of Climate Change

(And What You Can Really Do About it)

By Bill Wylson

Green Stem Press
Copyright © 2020

ISBN-13: 978-1-7342387-1-6
ISBN-10: 1-7342387-1-2

Green Stem Press
A White Horse Book

greenstempress.com
myldsbooks.com

"AND THERE SHALL BE A NEW EARTH."

ETHER 13:9

Table of Contents

Disclaimer:

Author's Note: Some of the ideas discussed in this work relate to doctrines and teachings of The Church of Jesus Christ of Latter-day Saints. However, the ideas expressed herein represent nothing more than the opinion of the author. I have no authority or commission to speak in any official capacity for the Church.

Publisher's Note:

Get More Out of This Book

Sterling Sill, author of over 30 books, once wrote about an article he read entitled *How to Get More Out of a Book Than There Is in It.* "Good readers," he explained, "may be able to get out of a book all there is in the book, but with a little imagination and some ability to analyze, they may get much more."

All capable readers can have their thoughts strike a particular notion, causing their thinking to drift away from the material they are reading. We should not be too quick to draw our minds back into the book, since frequently if we give our imagination a little freedom, it will direct us to some interrelated way of thinking that could prove to be extremely valuable.

People may often find that the most significant insights, ideas and beliefs are the ones that they come up with on their own and not so much from the concepts printed on the page. As our mind wanders along its own specific chain of correlated thought, we may arrive all on our own at some important interpretations and impressive conclusions. Then, when our minds have finished their journey of exploration and discovery, we can return our attention to the book and resume reading.

This is how to get more out of a book than there is in it. The book will cause us to come to conclusions regarding a diversity of notions not actually in the book. The interest of freeing our thoughts is an extremely beneficial and rewarding undertaking.

Paul, the New Testament apostle, was a known ponderer. He advises us that "whatsoever things are true, whatsoever things are honest, whatsoever things are just, whatsoever things are pure, whatsoever things are lovely, whatsoever things are of good report . . . think on these things." The ability to ponder gives us the capacity to obtain more from our circumstances and situations than what is actually in them. Through this procedure we place ourselves above the conventional and commonplace existence.

Thousands of fantastic, fascinating philosophies are frittering away in countless books. Hundreds of important and profound programs that could benefit us immensely sit untouched on library shelves. Even the word of God Himself remains largely unfamiliar and unacquainted to many of us. All the essential ingredients for success in any of our personal pursuits cannot advance our progression until we ingest and absorb them; until we get them circulating in our bloodstream and make them a part of our inner strength and learning.

As you read this book, or any other, practice the art of pondering. It will give you a more prolific passion for learning and thinking and, hopefully, for putting into practice. If what you read here does not please and persuade you, so much the better. You can amend each page or each chapter to your own specific situation to satisfy your own particular prerequisites.

Effective pondering will enable you to draw concrete conclusions and form compelling objectives on the vital subject of your personal progress in life.

Chapter One

Our Ever-Changing Earth

Climate change is real.

There really should be no question about it. However, polarized views about climate issues stretch from the causes and cures for climate change to issues of trust or skepticism in climate scientists and their research.

According to NASA: "Climate change is one of the most complex issues facing us today. It involves many dimensions — science, economics, society, politics and moral and ethical questions — and is a global problem, felt on local scales, that will be around for decades and centuries to come." [1]

---

[1] From NASA's Global Climate Change Website.

Here are some of the facts: [2]

The global temperature has increased 1.62 degrees (F) during the late 19th century.

Ocean surface temperatures have increased 0.4 degrees (F) since 1969.

The Greenland and Antarctic ice sheets have decreased in mass.

Glaciers are retreating almost everywhere around the world.

The amount of spring snow cover in the Northern Hemisphere has decreased over the past five decades.

Global sea levels rose about 8 inches in the last century.

The number of record high temperature events has been increasing, while the number of record low temperature events has been decreasing, since 1950.

---

[2] From NASA's Global Climate Change Website.

Here are some other facts:

The Earth's atmosphere is an active and dynamic fluid. It is constantly in motion. The atmosphere's physical properties as well as its rate and direction of motion are affected and influenced by a multiplicity of circumstances and considerations. These factors include solar radiation, ocean currents, the geographic position of continents, atmospheric chemistry, the location and orientation of mountain ranges, and the amount of vegetation growing on the Earth's surfaces. These underlying forces all change with time.

The Earth's climate results from physical properties and the motion of its atmosphere. Some of the dynamics of our climate can and will change within relatively brief timespans. These would include the distribution of heat within the oceans, atmospheric chemistry, and surface vegetation. Others, obviously, take longer, such as the position of continents and the location and height of mountain ranges. Climate change occurs at every imaginable interval.

As early as the 1800s, geologists detected indications of massive climatic changes which took place approximately 2.6 million years ago before the Pleistocene era. Today, the perpetual changes occurring with Earth's volatile climate are observed, examined and scrutinized by networks of sensors in space, on the surface of the Earth itself, and both on and below the Earth's oceans.

Over the past 200–300 years, especially since the early 1900s, climate change has been documented using instrumental records and other archives. Some written documents and other recordings make information available about climate change in certain localities for the past several hundred years. Certain chronicles indicating climate change have been identified dating back over 1,000 years, although these are very rare.

The only real question concerning climate change is: What can we do about it?

I have heard all kinds of solutions, from the bizarre to the ridiculous. The most recent suggestion I came across to reduce global warming was for us to stop ironing our clothes. Other suggestions I have noted have been to paint our rooftops white to reduce absorption of heat; to make highways white with black stripes instead of black with white stripes; to be vegan, thereby reducing the need for herds of animals that emit methane gases into the atmosphere; to not idle our vehicles for more than two minutes; and to turn our thermostats to 68 degrees (F).

Other suggestions have been to create a giant umbrella that could block two to four percent of the sun's rays and launch it into space; creating plastic trees that could absorb carbon dioxide; eating insects instead of meat; bio-engineering future humans to be smaller, smarter and with an intolerance to red meat; mass female sterilization; and, of course, we can't forget the suggestion from NASA scientists to move the earth farther away from the sun.

Don't misunderstand me. I live on this planet, too. I want cleaner air and purer water myself. I appreciate that not idling my vehicle can definitely help clean the air, but I also seriously question whether or not it will do anything to prevent climate change.

Why?

Because the same scientists who tell me that I should be wearing wrinkled clothes or eating a beetle instead of sirloin also tell me that climate change has been occurring on this planet since before one thousand million years ago. That was a time long before people were idling their SUVs and setting their thermostats too high. In fact, it was a time long before people.

The Earth's climate has been relatively stable for at least a couple of centuries now. Due to this comparative stability, cities and farms could be located in promising, beneficial climates without concern over whether the climate would change. But guess what. Earth's climate has continually changed throughout its entire history. A stable climate is simply not the norm.

Science has shown us that climate change has been a constant and continuing issue since the beginning of Earth's history. For a large portion of that history, the Earth has been hotter and more humid than it is at the present time. Conversely, the Earth has also seen periods of extreme cold, as when glaciers covered much of its surface.

According to NASA:

"The Earth's climate has changed throughout history. Just in the last 650,000 years there have been seven cycles of glacial advance and retreat, with the abrupt end of the last ice age about 11,700 years ago... Most of these climate changes are attributed to very small variations in Earth's orbit that change the amount of solar energy our planet receives." [3]

Numerous processes are involved in creating climate change. The amount of energy the Sun generates over time contributes extensively to climate change; the shift in the position of the continents would also be a contributing factor in climate change as would the tilt of Earth's axis; climate change has even been created by sudden and dramatic alterations in the Earth's orbit, triggered by sizeable asteroid impacts; and, of course, by greenhouse gases released into the atmosphere through natural occurrences or by human behaviors.

650 million years ago, for example, the earth's surface became entirely frozen; rainforests collapsed 305 million years ago; 55 million years ago global warming increased on average 41 to 46 degrees (F). All of this happened without humans burning fossil fuels, herding cattle or destroying the Brazilian rainforest.

The Earth's climate is influenced, affected and changed by a wide variety of circumstances

---

[3] From NASA's Global Climate Change Website.

that function on timetables ranging from fleeting hours to hundreds of millions of unending years. Countless causes of climate change are external to the Earth itself. Many other causes are part of the Earth's structure but are deemed external to the atmosphere. And then there are those causes that seem to suggest interactions between the atmosphere and other elements of the Earth system. Solar variability, volcanic activity, tectonic activity, orbital variations, among other recurring events, all contribute to changes in the Earth's climate.

Irrespective of where we make our home on this planet, all of us will, to some degree or another, experience changes within Earth's climate during our lifetimes. This is an inevitable fact. The most common and certainly the most foreseeable changes we will experience are the predictable phenomena known as seasonal cycles. Signs of seasonal changes include people either putting on more clothing or taking off extra clothing, adjusting thermostats either up or down, changing outdoor activities from picnicking to skiing as well as changes in agricultural practices such as planting or harvesting.

Even with these common, seasonal changes no two summers and no two winters are ever alike in the same location. Some summers are hotter, some are rainier, some are drier. We can blame rises in fuel costs, lower crop yields, increased road maintenance budgets, and hazardous wildfires on these interannual disparities in the weather.

These same variations also account for single-year, precipitation-driven flooding like the 1993 flood in the Mississippi River basin or the devastating and deadly floods in Bangladesh in 1998. The more recent wildfires in California and Australia, severe storms and hurricanes like Dorian that only a short time ago demolished the Bahamas, and other climate-related events are all the result of interannual climate variation.

In addition to seasonal variations, science presents evidence for Interannual variation, Decadal variation, Centennial-scale variation, Millennial and Multi-millennial variation. Some changes in Earth's climate occur over decades. Certain localities have seen multiple years of drought while others have experienced severe flooding or similar unhospitable occurrences. Decadal differences in climate can disrupt water supplies triggering crop failures and other disasters. The dust bowl droughts in the United States in the 1930s are an observable example of decadal climate change. Such changes can cause widespread food shortages similar to what occurred during the Sahel drought in northern Africa in the 1970s and '80s.

According to the scientists, the Earth has experienced and endured a universal cooling shift for the past 50 million years. This trend culminated with the development of permanent ice sheets in the Northern Hemisphere. Over millennia, as the ice sheets waxed and waned, global climate drifted steadily toward cooler conditions characterized by increasingly severe glaciations and increasingly cool interglacial phases.

These phases are caused ultimately by two mechanisms of Earth's orbital geometry:

1) the equinoctial precession cycle
2) and the axial-tilt cycle.

And yet, despite all of these scientific indications, for some reason scientists want to attribute our current climate change to the human expansion of the "greenhouse effect" — warming that results when the atmosphere traps heat radiating from Earth toward space.

Chapter Two

Politics behind Climate Change

"Science is not a body of indisputable and immutable truth." [4]

Former White House Advisor Rahm Emanuel is credited with saying: "You never want a serious crisis to go to waste." When dealing with a significant   disaster, opportunistic politicians

---

[4] Standen, Anthony, *Science is a Sacred Cow*, E.P. Dutton, 1950.

will never miss an occasion to convert crises into a political gain for special interests.

In other words, key crises offer a guise for allocating and dispensing benefits to targeted special interest groups. The bigger the crisis, the larger the paybacks, profits and compensations that get packed into the final legislation. This creates an absolute guarantee that any effort to resolve the crisis will be delayed and postponed until the very last possible moment.

This, of course, is true of the so-called climate crisis of today. The political disagreements over climate change extend far beyond simply whether climate change is real or whether humans are even a contributing factor. These distinctions reach across every dimension of the climate change debate. Even our basic trust in the motives behind the climate scientists' research and findings is threatened by unprincipled and devious policymakers.

Vast differences exist in political opinion as to the potential for devastation brought on by climate change and the manner in which to contend with climate change impacts. The differing levels of trust in the information obtained from scientific researchers, vastly divergent perceptions of scientific consensus, and differing views as to whether climate scientists are driven by a quest for knowledge or a quest for professional advancement cloud the real issues with uncertainty, reservation and confusion.

Unquestionably, the biggest differences in climate science and climate policy are found between the opposite ends of the political spectrum. Whether we are talking about probable causes or possible solutions, liberal Democrats and conservative Republicans see what is happening to Earth's climate through vastly differing lenses.

Liberal Democrats demonstrate greater trust in climate scientists' research and understanding of the situation. Therefore, liberal Democrats are more inclined to believe that a wide variety of natural disasters and environmental catastrophes are potentially on our doorstep. (And they certainly are!) The liberal Democrats are also more inclined to believe that governmental policies and personal carbon footprint reduction can aid in impeding these potential disasters. (Which they certainly won't!)

Republicans, on the other hand, even those who accept climate change as a reality, appear less likely to envisage severe damage to our planet or to believe that any individual actions or political policies would make a dent in deterring disasters. At the very least, they may be willing to admit that certain changes can make a small albeit insignificant difference.

As of 2019, the political division over climate change looked something like this:

- *Power plant emission restrictions:* 76% of liberal Democrats say this can make a big difference, while only 29% of conservative Republicans say the same, a difference of 47 percentage points.
- *An international agreement to limit carbon emissions:* 71% of liberal Democrats and 27% of conservative Republicans say this can make a big difference, a gap of 44 percentage points.
- *Tougher fuel efficiency standards for cars and trucks:* 67% of liberal Democrats and 27% of conservative Republicans say this can make a big difference, a 40 percentage-point divide.
- *Corporate tax incentives to encourage businesses to reduce the "carbon footprint" from their activities:* 67% of liberal Democrats say this can make a big difference, while only 23% of

conservative Republicans agree, a difference of 44 percentage points.

- *More people driving hybrid and electric vehicles:* 56% of liberal Democrats say this can make a big difference, while 23% of conservative Republicans do, a difference of 33 percentage points.
- *People's individual efforts to reduce their "carbon footprints" as they go about their daily lives:* 52% of liberal Democrats say this can make a big difference compared with 21% of conservative Republicans, a difference of 31 percentage points. [5]

Moderate-liberal Republicans and moderate-conservative Democrats vacillate somewhere in between the illogical and ideological ends of either of the more extremist groups.

The American poet, Ogden Nash, brought to light a solemn truth when he wrote: "I believe that people believe what they believe they believe." [6] In other words, if someone refuses to accept a statement about one thing or another, their refusal isn't based on whether that one thing is true or not. It is, rather, based on

---

[5] Pewreseach.org.
[6] Nash, Ogden, *Good Intentions*, Doubleday, p.100.

whether that one thing agrees with what they already believe to be true. Likewise, when they accept something as a truth, it isn't based on the truth itself but on whether it agrees with an idea they have already accepted as true.

Most people accept scientific evidence as fact even though many of the so-called "scientific facts" have now been proven incorrect, inaccurate or flawed. "Science is *not* a body of indisputable and immutable truth," writes Anthony Standen. "It is a body of well-supported probable opinion only, and its ideas may be exploded at any time." [7] Scientific method entails supposition and conjecture, and these always encompass the probability of error. One important factor to keep in mind is that scientists also have a political agenda.

When our opinions are based on false assumptions, we will often defend all kinds of mendacities as true and accurate. As Ogden Nash eloquently explains: "The door of a bigoted mind opens outward so that the only result of the pressure of facts upon it is to close it more snugly." [8]

---

[7] Standen, Anthony, *Science is a Sacred Cow*, E.P. Dutton, 1950.
[8] Nash, Ogden, *Good Intentions*, Doubleday, p.100.

One of the basic assumptions inherent in today's climate change model is the assumption that carbon dioxide and other greenhouse gases are the root cause of rising temperatures. This may be true; however, numerous scientific studies demonstrate that the climate change models have grossly overstated the amount of warming the Earth has experienced as carbon dioxide levels have increased. Scientific research has demonstrated that for 522 million years, atmospheric carbon dioxide levels have had no causal relation to temperature or climate conditions.

So, why all the emphasis on fossil fuels and gas-guzzling SUVs?

If you were to attend a social event and you were told that one of the guests was recently released from prison where he served five years for robbery, and if your wallet or purse suddenly turned up missing, who would you immediately suspect? Most likely you would think it was the man you were told was a thief. You might not even consider how close the person who told you about the so-called thief was standing next to you when he told you this.

The majority of the information we receive regarding climate change is being fed to us by the Intergovernmental Panel on Climate

Change or the IPCC. The truth about the history of climate change and the politics surrounding climate change is being largely hidden from the world's population.

The IPCC, since its inception, has been a political organization and not a scientific organization. Its political directive always has been to research the human cause of climate change. Consider that for a moment in the context of the thief I mentioned earlier. The idea that climate change is human caused is already built into their directive! Why even bother to look at natural causes? Why study the role of the Sun or the cycles of the ocean or historic causes of climate change? Their quest is to "discover" the human cause of climate change. They have a political agenda that must be satisfied. Their findings are formulated by politicians, not scientists, to satiate their desires for even greater governmental control over the people and the economy and increased power to deliver funding (i.e., increase taxes on the public) to governmental bureaucracies and other non-governmental organizations. Significant scientific uncertainties regarding the causes and consequences of the present period of climate change are often downplayed or outright ignored in order to further the IPCC's intended agenda.

The IPCC can't even make accurate basic temperature predictions. The IPCC's temperature projections failed to match recorded temperatures, thereby throwing doubt on their other projections regarding disease, drought, extinctions, famine, flood, hurricanes, and sea levels.

I'm not saying that climate change is a hoax. Right from the very beginning of this book I stated that climate change is real. But is there a hoax involved? There certainly is!

The climate is changing. It always has changed. You can travel to Greenland and see the massive chunks of glacial ice falling into the ocean causing sea levels to rise to what could become dangerous levels. You can study scientific reports that will claim human causes are behind these changes, but other scientific reports will tell you the exact opposite.

Scott Morrison, Australia's Prime Minister, concedes that climate change has had an influence on the Australian wildfires, and he has championed his government's climate record, but even the Prime Minister acknowledges that governmental climate change bureaucracies with all of their "job-destroying, economy-destroying, economy-wrecking targets and goals won't change the fact that there have been bushfires or anything like that in Australia."

Australian researcher John McLean has uncovered discrepancies in the datasets used by the IPCC in formulating its climate projections. McLean exposed various data variances, implausible and far-fetched data claims, and outright errors in the dataset including evidence to manipulate recorded temperature readings. McLean also found those compiling the dataset consistently misrepresented or "altered" recorded temperatures in a way to make temperatures appear cooler in the earlier part of the twentieth century and warmer in the latter part, thereby amplifying the rapidity and extent of the Earth's warming.

You've probably heard the statement that "the only constant in life is change." Well, this is also true of Earth's climate; it has constantly been changing, according to scientists, over millions of years—with and without human interaction.

Hurricanes are getting worse. Wildfires are demolishing entire neighborhoods, even cities. Earthquakes are increasing in frequency and force. All of these are of grave concern to people all over the Earth.

The Earth's climate has been unstable throughout history. Whether our current climate change is the result of human behaviors, natural influences, or an amalgamation of both is a widely debated question. But there is no question that politics, including the thirst for increased funding and the lust for power, rather than a quest for truth and knowledge, is the motivation behind the IPCC's efforts.

Of one thing we can be *absolutely* certain: Raising taxes will not stop climate change!

Chapter Three

A Pained and Weary Planet

"Wo, wo is me, the mother of men; I am pained, I am weary." [9]

Until we can fully appreciate the immeasurable and interminable importance of the Earth and accept our tenacious relationship to it, we will continue to be afflicted and inundated with worldwide instability, upheaval and distress. In our extensive search for solutions to climate change we are overlooking

---

[9] Moses 7:48.

the one possibility that would make a real and significant difference. It is vital, then, that we examine and understand historic events that have shaped, altered and transformed our planet.

Everything formed and created, including the Earth, was fashioned for a purpose. Everything that complies with its purpose will be advanced in a never-ending, infinite process of improvement.

Until the early 1800s, the common scientific consensus concerning the creation and history of our planet was that it had existed as molten rock and that during a cooling process immense cataclysmic events took place. These earth-shattering incidents included massive volcanic eruptions as well as sudden and violent uplifts of mountain ranges. Evidence also exists indicating the impact of large meteorites that pummeled the Earth, causing major changes in its surface.

The collective thought, even among many of the scientists of that period, was that the Earth's history coincided with the biblical timeline of creation. Scientific religionists tended to relate calamitous occurrences, such as the worldwide flood of 2348 BC, to the biblical narrative. Most scientists and theologians subscribed to the opinion that the Earth was formed as a habitable planet around 4000 BC.

The concept of catastrophism, as this perspective was called, differs from the concept of uniformitarianism put forth by British lawyer turned geologist, Charles Lyell, in the latter part of the 1800s. The basic presupposition separating the theory of uniformitarianism from the premise of catastrophism is the assumption that Earth had a benign history and was formed over millions of years through the slow processes of erosion and mountain formation occurring consistently over time at the same rate that they are observed to occur today.

Lyell expanded on a theory formulated by a Scottish farmer, James Hutton, some fifty years earlier, who argued that the earth was not changed by sudden catastrophic events, but by unperceptively slow changes, such as those that are observed today. Hutton suggested that the Earth's surface was formed through the gradual uplift of mountains from Earth's molten core and the rain-erosion process that played upon those mountains. These extremely slow, steady elevations and consequent erosions repeated and continued over extensive time periods would create a type of perpetual motion that would shift in recurring cycles of wearing away and reforming until the planet finally became suitable for the occurrence of humanity. This version of geology was called uniformitarianism, because of Lyell's fierce insistence that the processes that alter the earth are uniform over time.

In April, 1940, a Russian scientist by the name of Velikovsky suggested that a great natural catastrophe might have taken place on Earth at the time of the Israelites' Exodus from Egypt that could account for the plagues that occurred, the parting of the Sea of Passage, the eruption on Mount Sinai, and the pillar of cloud and fire that moved in the sky. But Velikovsky needed proof outside of the biblical account that could verify this supposition. An obscure papyrus stored in Leiden, Holland held the evidence Velikovsky was seeking. The Ipuwer Document containing the lamentations of the Egyptian sage, Ipuwer parallels the Book of Exodus and describes the same natural catastrophes and the same plagues as the Old Testament account. Velikovsky synchronized the histories of Egypt and Israel, taking this catastrophe — which brought the downfall of the Egyptian Middle Kingdom — as a starting point.

About six months later, in October of 1940, Velikovsky identified another important fact. He noticed that Joshua details a destructive shower of meteorites [10] which took place just prior to the time when the sun "stood still" in the sky. The Book of Joshua tells us that "the Lord cast down great stones from heaven upon them." [11] The Book of Jasher gives the following account:

"The Lord sent upon them hailstones from heaven, and more of them died by the hailstones, than by the slaughter of the children of Israel...

"And when they were smiting, the day was declining toward evening, and Joshua said in the sight of all the people, Sun, stand thou still upon Gibeon, and thou moon in the valley of Ajalon, until the nation shall have revenged itself upon its enemies.

"And the Lord harkened to the voice of Joshua, and the sun stood still in the midst of the heavens, and it stood still six and thirty moments, and the moon also stood still and hastened not to go down a whole day.

---

[10] Joshua 10:11.
[11] Joshua 10:13.

And there was no day like that, before it or after it." [12]

This was not mere coincidence. The ancient prophets were recording a cosmic disturbance that certainly would have shaken the entire Earth. Velikovsky speculated on whether these events could be related to the upheavals and plagues which occurred approximately fifty years earlier during the Exodus from Egypt. After surveying other sources from around the globe, Velikovsky became convinced that a worldwide disaster had overtaken the Earth, and that the planet Venus played a decisive role in that cataclysm. For the following ten years, Velikovsky researched and wrote *Ages in Chaos* and *Worlds in Collision*.

Following in brief are some of the conclusions of Velikovsky's work that have so disturbed the proponents of uniformitarianism:

1. "The Erratic Venusian Orbit.

"Velikovsky discovered that the sun-earth-moon system has undergone two periods

---

[12] Jasher 88:61-65.

of extreme disturbance which have changed the parameters of the system by large amounts within the past few thousand years. In the first period, the earth had two near collisions with Venus, the first being in the time of the Hebrew exodus from Egypt and the second being on the day during Joshua's campaigns when the sun and moon stood still. These near collisions were about 50 years apart, and they took place sometime around 1500 BC. Velikovsky points to literature of the period that refers to the 'horns' of Venus, an obvious reference to the crescent planet, which came near enough to the earth to be seen as a sphere rather than as a point of light, and since it is nearer to the sun than is the earth, would be seen in phases, as is the moon.

2. "The Erratic Martian Orbit

"During the second period, the earth had two or more near collisions with Mars, and Velikovsky gives exact dates for the first and last of them. The first was on February 26, 746 BC and the last was on March 23, 686 BC. Since March 23, 686 BC the parameters of the sun, moon, and planets have been undisturbed. For at least part of the time between 746 and 686 BC, the month had 36 days while the year continued to have 360 days, according to Velikovsky. Thus, he explains the fact that the oldest known Roman calendar had only 10 named months.

"The names of the last four of the ten months of the year were September, October, November and December indicating the Seventh, Eighth, Ninth and Tenth months. Sometime before 686 BC the continued disruptions of the lunar orbit by near-passes of the planet Mars resulted in a change of the orbital period from 36 to 29.53 days. This caused so much trouble with the calendar of 36-day months that in 45 BC Julius Caesar commissioned the 12-month calendar to be developed. To the 10-month calendar they added in the middle of the year two new months-July and August (Julius and Augustus). This Julian calendar held sway until 1582, when Pope Gregory XIII instituted a 'new and improved' calendar (called the Gregorian calendar) that replaced the older Julian calendar. The new calendar was quickly adopted in most Catholic countries, but the many Protestant and Orthodox countries continued to use the older Julian calendar for centuries. Russia, for instance, continued to use the Julian calendar until Feb 14, 1918.

"Velikovsky also predicted that Venus had retrograde rotation, a heavy hydrocarbon atmosphere and a surface temperature on the order of 800^0 C. All of those predictions were verified by subsequent space probes of Venus conducted by NASA." [13]

Indisputable evidence exists, in addition to Velikovsky's findings, supporting the fact that Earth's history has been anything but benign. Even the proponents of uniformitarianism agree that cataclysmic events brought about the demise of the dinosaurs. They seem to agree that the impact of a large meteorite approximately 100 miles wide hitting the Yucatan coastline resulted in the extinction of these enormous reptiles. The heat impact of such an event would have created a fireball and searing vapor cloud which would have set the North American continent on fire. The environmental aftereffects led to the global extinction of many plants and animals, including the dinosaurs.

There exists an important underlying parallel between our planet and its human population. An accurate and correct perception

[13] Gorton, H. Clay, *The Demise of Darwinism*, White Horse Books.

of this correlation is crucial in understanding current climate change conditions. Our Earth had an auspicious and providential beginning. It was a paradise. The first people to inhabit the Earth, Adam and Eve, resided on this paradisiacal planet and lived in a way that allowed them to remain in the presence of God.

But Adam fell. This was not unforeseen or unexpected. It was a crucial part of the Creator's plan. However, with Adam's fall the Earth also fell and was cursed. The ground itself became cursed and began to produce thorns and thistles. Adam and all his descendants, the entire human race that followed, were exiled from the presence of God. The Earth and its human inhabitants now both live with the assurance of an unimaginable, magnificent recovery and regeneration. A potential new life for both the planet and its people is at our doorstep.

Our past experience on this planet has confirmed and proven, time and time again, that moral compliance to the will of God will always bring about a beneficial and advantageous cooperation from the Earth's elements. Climate can essentially change to our advantage or disadvantage, depending on how well we comply to the edicts of God.

Enoch, a pre-diluvian prophet,

understood and acknowledged the close connection that exists between the climate and productivity of the Earth and the principled obedience of its inhabitants. Enoch perceived groans and lamentations rising out of the Earth because of the malice and iniquity of its people.

"Enoch looked upon the earth; and he heard a voice from the bowels thereof, saying: Wo, wo is me, the mother of men; I am pained, I am weary, because of the wickedness of my children. When shall I rest, and be cleansed from the filthiness which is gone forth out of me? When will my Creator sanctify me, that I may rest, and righteousness for a season abide upon my face?" [14]

After the fall of Adam and Eve (and the descent of humanity which ensued,) a divine verdict proclaimed that: "…cursed is the ground for thy sake … thorns and thistles shall it bring forth unto thee … and in the sweat of thy face shalt thou eat bread till thou return unto the ground." [15]

Regardless of the declaration to Adam, God promised ancient Israel that if the people would be faithful and live righteously, the

---

[14] Moses 7:48.
[15] Genesis 3:17-19.

climate and seasons would operate to their advantage.

"If ye shall harken diligently unto my commandments which I command you this day…

"…I will give you the rain of your land in his due season, the first rain and the latter rain, that thou mayest gather in thy corn, and thy wine, and thine oil." [16]

In other words, the ripening rains were a consequence of their obedient nature as was the bountiful harvest to follow. However, they were also warned to not be deceived and turn aside, otherwise God would "shut up the heaven, that there be no rain, and that the land yield not her fruit." [17]

A similar promise was made to Alma, an ancient prophet of the Americas:

"Blessed are thou and thy children; and they shall be blessed, inasmuch as they shall keep my commandments they shall prosper in the land." [18]

---

[16] Deuteronomy 11:13, 14.
[17] Deuteronomy 11:16,17.

Simply stated, our good behavior makes the soil more fertile and the Earth itself more productive. Conversely, history demonstrates that our sinful behaviors are the cause of many of the natural calamities and changes in climate conditions that transpire on our planet. Earth's natural disasters are triggered by immoral and irreligious conduct.

The tiny country of Palestine, during the time of Israel's moral rectitude and decency, could sustain and support an exceptionally substantial populace. However, this once substantial land became little more than a desert, not because of single-use plastic straws or carbon emissions or an overabundance of livestock methane gases; it was the direct result of the disobedience of the people toward the laws of God. The Earth is degraded and polluted, not only by carbon emissions and indestructible garbage bags, but also by the misconduct of its human inhabitants. The Lord told Isaiah: "The earth also is defiled under the inhabitants thereof because they have transgressed the laws, changed the ordinances, and broken the everlasting covenant." [19]

---

[18] Alma 50:20.
[19] Isaiah 24:5.

Under the rule of Ahab and Jezebel, the worship of Baal and Asherah was established in Northern Israel and attempts were made to kill the prophets of God and end the worship of Jehovah. [20] The prophet Elijah sealed the heavens and rain was withheld from Israel for approximately three years. [21]

Babylon, once an illustrious world center, grew desolate because of the depravity, iniquity and immorality of its inhabitants. In Noah's time the debauchery and indulgences of Earth's population opened the flood gates of heaven and caused the fountains of the great deep to release a great flood of waters upon the earth. [22] The indulgences and transgressions of the residents of Sodom and Gomorrah brought fire and brimstone down from the heavens. [23] And yet, if there had been just ten good people living in those cities, the citizens of Sodom and Gomorrah would have been spared. Their personal conduct toward God's laws could have altered the natural disaster that destroyed them.

The impiety of the people in the days of

---

[20] 1 Kings 16:32,33 and 18:13.
[21] 1 Kings 17:1.
[22] See Genesis 7.
[23] See Genesis 19.

Jesus plunged our planet into the lengthy blackness of the dark ages where mental and spiritual advancement came to an abrupt halt. When Christ was crucified on Calvary darkness fell and "the earth did quake and the rocks were rent." [24]

Widespread and boundless destruction and devastation also accompanied the Savior's death on the American continents. Dreadful storms with thundering and lightning, and great earthquakes caused a magnitude of unprecedented destruction which transformed the Earth's surface. The convulsions of the Earth brought down the mountains; valleys were raised up and numerous cities were destroyed. Further destruction was caused by fire as well as by the sea flowing into the newly formed valleys. The inhabitants of the Americas suffered the destructive forces of nature because they flouted prophetic warnings and failed to repent.

These destructive occurrences and divine punishments do not have natural but rather supernatural causes. They are superhumanly regulated. Destructive climate changes innately and inevitably follow the turpitudes of humanity

---

[24] Matthew 27:52.

and the degenerate status of our race.

Conversely, numerous illustrations exist of considerable periods of prosperity that are the direct result of righteous behavior by the people on Earth. One such era is represented as follows:

"And it came to pass in the thirty and sixth year, the people were all converted unto the Lord upon all the face of the land... and there were no contentions or disputations among them; and every man did deal justly one with another.

"...there were not rich and poor, bond and free but they were all made free, and all were partakers of the heavenly gift.

"...and the Lord did prosper them exceedingly in the land." [25]

Prior to Noah, Enoch proclaimed repentance to a reprobate and degenerate people. Enoch attained such as state of holiness that he walked with God. [26] Because of his righteous standing, at Enoch's command the Earth trembled, and the mountains fled. This mighty man of God built a city of such virtuous

---

[25] 4 Nephi 1:2,3,7.
[26] See Moses 6:34.

and upright inhabitants that the entire population was taken up to God. [27]

During the reign of Jeroboam II, about 750 years before Christ, God sent Jonah to the city of Nineveh because "their wickedness is come up before me." [28] Jonah prophesied to Nineveh's populations of an impending destruction confronting them. Because of his warning, the people repented. The King actually ordered everyone to turn "from his evil way and from the violence" that was in their hands. "Who can tell if we will repent, and turn unto God, but he will turn away from us his fierce anger that we perish not?" proclaimed their king. "And God saw their works that they turned from their evil way and repented, and God turned away the evil that he said he would bring upon them." [29]

Two fascinating and curious extremes are positioned directly before our own world today. It has been predicted that we will go through horrifically intense catastrophes and natural disasters prior to the second coming of Christ. These natural calamities will precede the millennial reign when the Earth will experience a thousand years of peace. We have the power

---

[27] Genesis 5:24; Helaman 11:5.
[28] Jonah 1:2.
[29] Jonah 3:7-10. (JST)

by turning to God to lessen the Earth's curse and sanctify its ground. Like the citizens of the City of Enoch and like the people of Nineveh, we can choose to hallow our homes, sanctify our families and consecrate our objectives by turning from our wrongdoings and pledging our lives to God and to virtue, righteousness and morality.

Chapter Four

What Tomorrow Holds

"All you now know can scarcely be called a preface to the sermon that will be preached with fire and sword, tempests, earthquakes, hail, rain, thunders and lightnings, and fearful destruction." [30]

The Old Testament prophet, Isaiah, foresaw the natural disasters and environmental catastrophes facing our present world. He warned us that we would be reprimanded and

---

[30] Young, Brigham, *Journal of Discourses* 8:123.

chastened with geographical and biological calamities because of our defiance toward God.

"Howl ye; for the day of the Lord is at hand; it shall come as a destruction from the Almighty.

"Therefore shall all hands be faint, and every man's heart shall melt:

"And they shall be afraid: pangs and sorrows shall take hold of them; …

"Behold, the day of the Lord cometh, cruel both with wrath and fierce anger, to lay the land desolate: and he shall destroy the sinners thereof out of it.

"For the stars of heaven and the constellations thereof shall not give their light: the sun shall be darkened in his going forth, and the moon shall not cause her light to shine.

"And I will punish the world for their evil, and the wicked for their iniquity; I will cause the arrogancy of the proud to cease, and will lay low the haughtiness of the terrible.

"Therefore I will shake the heavens, and the earth shall remove out of her place, in the wrath of the Lord of hosts, and in the day of his fierce anger …

"And Babylon, the glory of kingdoms, the

beauty of the Chaldees' excellency, shall be as when God overthrew Sodom and Gomorrah." [31]

Concerning this impending devastation, the Lord revealed to the modern-day prophet, Joseph Smith that: "not many days hence and the earth shall tremble and reel to and fro as a drunken man; and the sun shall hide his face, and shall refuse to give light; and the moon shall be bathed in blood; and the stars shall become exceedingly angry, and shall cast themselves down as a fig that falleth from off a fig tree.

"And after your testimony cometh wrath and indignation upon the people.

"For after your testimony cometh the testimony of earthquakes, that shall cause groanings in the midst of her, and men shall fall upon the ground and shall not be able to stand.

"And also cometh the testimony of the voice of thunderings, and the voice of lightnings, and the voice of tempests, and the voice of the waves of the sea heaving themselves beyond their bounds.

"And all things shall be in commotion; and surely, men's hearts shall fail them; for fear shall come upon all the people." [32]

---

[31] See Isaiah 13.

The ancient American prophet Nephi envisaged the iniquity and evilness that characterizes the world today. His prophecy pertains both to those "who shall come upon this land and those who shall be upon other lands." and will be fulfilled when we are "drunken with iniquity and all manner of abominations." He then predicted the judgments of God that we would experience when we "shall be visited of the Lord of Hosts, with thunder and with earthquake, and with a great noise, and with storm, and with tempest, and with the flame of devouring fire." [33]

These predicted changes in our climate and in our environment are not the result of carbon emissions but of compliance omissions. They are the direct consequence of our failure to follow the word and the will of God. It isn't taxation or IPCC programs that will save the Earth but the observance of obedience to the Lord's commandments.

A revelation given in 1833 states: "The Lord's scourge shall pass over by night and by day, and the report thereof shall vex all people; yea, it shall not be stayed until the Lord come;

---

[32] D&C 88:87-91.
[33] 2 Nephi 27:1,2.

"For the indignation of the Lord is kindled against their abominations and all their wicked works.

"Nevertheless, Zion shall escape if she observe to do all things whatsoever I have commanded her.

"But if she observe not to do whatsoever I have commanded her, I will visit her according to all of her works, with sore affliction, with pestilence, with plague, with sword, with vengeance, with devouring fire." [34]

Brigham Young also warned us about calamities which nature would unleash on the world:

"All you now know can scarcely be called a preface to the sermon that will be preached with fire and sword, tempests, earthquakes, hail, rain, thunders and lightnings, and fearful destruction. ... You will hear of magnificent cities, now idolized by the people, sinking in the earth, entombing the inhabitants. The sea will heave itself beyond its bounds, engulfing mighty cities. Famine will spread over the nations, and nation will rise up against nation, kingdom against kingdom, and states against states, in our

---

[34] D&C 97:23-26.

own country and in foreign lands; and they will destroy each other, caring not for the blood and lives of their neighbours, of their families, or for their own lives." [35]

Other prophesies warn of terrible diseases and sickness that will cover the Earth because we turned away from God. We are told that we will see "an overflowing scourge; for a desolating sickness shall cover the land." [36] This desolating scourge will "continue to be poured out from time to time, if [we] do not repent, until the earth is empty." [37]

Prophets have also written of an incredible hailstorm yet to come that will destroy the crops of the Earth. [38]

Prophecies have been made regarding various earthquakes and storms. The frequency and severity of earthquakes is increasing throughout the world. There will be at least three specific earthquakes with far-reaching effects. One of those earthquakes will take place in the United States of America. Prophets have foretold of an earthquake that will destroy New York City. Other predictions reveal that Albany,

---

[35] Journal of Discourses 8:123.
[36] D&C 45:31.
[37] D&C 5:19.
[38] See D&C 29:16.

New York will be devastated by fire and that Boston will be swept into the ocean, perhaps by a Tsunami or tidal wave. [39] It will be the same in numerous other cities. [40]

Through the Prophet Joseph Smith the Lord also revealed that "God has set his hand and seal to change the times and seasons, and to blind their minds, that they may not understand his marvellous workings; that he may prove them also and take them in their own craftiness." [41]

Drought will also play an important part in the coming climate changes on Earth: "Behold, at my rebuke I dry up the sea. I make the rivers a wilderness; their fish stink, and die for thirst.

"I closed the heavens with blackness, and make sackcloth their covering.

"And this shall you have at my hand — ye shall lie down in sorrow." [42]

This period of partial judgment and overwhelming climate change preceding the second coming of Christ will clear the way for the thousand years of peace when "Christ will

---

[39] Journal of Discourses 21:299; 12:344.
[40] Journal of Discourses 20:152.
[41] D&C 121:12.
[42] D&C 133:68-70.

reign personally upon the earth; and, ... the earth will be renewed and receive its paradisiacal glory." [43]

At that time, the virtue, morality and decency of humanity will abolish the curse which is on the Earth and our planet will experience a transformation and be regenerated. The elements will melt with fervent heat and the Earth will be consumed. A new Earth will materialize in the form of a replenished and revitalized planet.

[43] Articles of Faith 10.

Chapter Five

A New Earth

"The whole earth is at rest, and is quiet: they break forth in singing." [44]

The time is coming when the turmoil on our beautiful planet will cease and the Earth will rest. During the Millennial reign the people of the Earth will live peaceably with each other in a state of happiness and prosperity.

At this time the enmity between humans and animals will end. The battle between right

---

[44] Isaiah 14:7.

and wrong, between good and evil, will also come to an end. The final transition will occur as our current world will perish and all things will become new.

John the Revelator saw and wrote about this future condition:

"And I saw a new heaven and a new earth: for the first heaven and the first earth were passed away; and there was no more sea.

"And I John saw the holy city, new Jerusalem, coming down from God out of heaven, prepared as a bride adorned for her husband." [45]

The Prophet Ether confirms John's vision and writings:

"And there shall be a new heaven and a new earth; and they shall be like unto the old save the old hath passed away, and all things have become new." [46]

The written word of God depicts a comprehensive portrayal of the circumstances which will exist during the 1000-year timeframe known as the millennium. According to the prophet Isaiah, the millennium will be an era of

[45] Revelation 21:1,2.
[46] Ether 13:9.

peace. Hostilities in the animal world which we now consider as natural will come to an end. Any contention, rivalry or disorder will not be permitted to exist.

"The wolf also shall dwell with the lamb, and the leopard shall lie down with the kid; and the calf and the young lion and the fatling together; and a little child shall lead them.

"And the cow and the bear shall feed; their young ones shall lie down together: and the lion shall eat straw like the ox.

"And the sucking child shall play on the hole of the Asp, and the weaned child shall put his hand on the cockatrice' den.

"They shall not hurt nor destroy in all my holy mountain; for the earth shall be full of the knowledge of the Lord, as the waters cover the sea." [47]

This new Earth shall be elevated and exalted to its definitive position of the celestial Kingdom. The Earth will then be prepared and fitted for the presence of God. As John wrote:

"…And God himself shall be with them, and be their God." [48]

---

[47] Isaiah 11:6-9.
[48] John 21:3.

The home life of the inhabitants of this new Earth will continue during the Millennium much the same as it is now. According to Isaiah:

"And they shall build houses, and inhabit them; and they shall plant vineyards, and eat the fruit of them." [49]

Brigham Young envisioned a similar image of life during the Millennium:

"The Millennium consists in this — every heart in the church and Kingdom of God being united in one; the Kingdom increasing to the overcoming of everything opposed to the economy of heaven, and Satan being bound, and having a seal set upon him. All things else will be as they are now, we shall eat, drink, and wear clothing. Let the people be holy, and the earth under their feet will be holy." [50]

Since wickedness will no longer exist on the Earth, absolute joyfulness, contentment and prosperity will dominate.

"And God shall wipe away all tears from their eyes; and there shall be no more death, neither sorrow, nor crying, neither shall there be anymore pain: for the former things are passed

---

[49] Isaiah 65:21.
[50] Journal of Discourses 1:203.

away." [51]

And then, the Earth shall rest.

"And there shall be mine abode, and it shall be Zion, which shall come forth out of all the creations which I have made; and for the space of a thousand years the earth shall rest." [52]

"Let the people be holy, and the earth under their feet will be holy."

Brigham Young

---

[51] John 21:4.
[52] Moses: 7:64.

Climate Change and the Making of an American Prophet

Of all the people living on the earth in 1820, God chose Joseph Smith, an obscure, uneducated and insignificant farm boy to establish His Kingdom, restore His priesthood and translate His words recorded on plates of gold buried in the hill Cumorah. Was it just by some freak coincidence that the chosen prophet lived only a few miles from where the ancient Prophet Moroni had buried the golden plates 1400 years earlier? Or did the Lord have a hand in getting Joseph to the right place at the right time?

For three generations, Joseph Smith, Sr.'s ancestors were affluent and prominent residents of Topsfield, Massachusetts. Joseph Sr.'s grandfather, Samuel Smith, was a highly respected person in the village and was frequently elected town selectman, town clerk, and representative to the Massachusetts legislature. The Salem Gazette referred to him as an "esteemed... man of integrity and uprightness... a sincere friend to the liberties of his country, and a strenuous advocate for the doctrines of Christianity." Samuel died when the market for farm products was in a deep slump and farm debt plagued thousands of Massachusetts farmers. At the time of his death, Samuel's estate was insolvent. Samuel's son, Asael, took over the family farm for about five years, until he too failed and was forced to sell the property and move. Asael's two oldest sons went to Vermont to a new town called Tunbridge. There they cut trees through the summer of 1791 and built a tiny hut. In the fall the entire family arrived and crowded into the little shelter to weather the winter.

Joseph Smith, Sr. met Lucy Mack through one of Lucy's brothers, Stephen Mack and they married in 1796. Joseph's father Asael gave his son a farm, and Stephen gave Lucy a thousand dollars, a substantial sum of money at that time.

But they did not enjoy their riches for long. In 1802 they rented out the farm and opened a store in Randolph, a neighboring town. Unfortunately, the store failed, leaving the Smiths with heavy debts. They sold their farm and dispensed all of Lucy's thousand-dollar wedding gift just to meet their obligations. They were left with nothing.

Joseph, Sr. and Lucy Smith, once established property holders, abruptly fell to the status of tenant farmers. They went from farm to farm in towns along the Connecticut River, crossing from Vermont into New Hampshire and back to Norwich.

Then it hit!

The year was 1816 and is known as the Year Without a Summer, the Summer that Never Was, the Poverty Year, or as Vermont folk singer Pete Sutherland referred to it in his song; "1800 and Froze-to-Death."

Severe summer climate abnormalities caused average global temperatures to decrease by 0.7 to 1.3°F. Evidence suggests this variance was caused by a historic low in solar activity combined with a volcanic winter created by a series of major volcanic eruptions. This included the 1815 eruption of Mount Tambora, in Indonesia, the largest known eruption in over 1,300 years. It was the largest eruption since the Hatepe eruption in 180 AD and it occurred during the middle of the Dalton Minimum (a period of unusually low solar activity).

The effects were widespread and lasted beyond the winter. At the time, Europe was still recovering from the Napoleonic Wars and had experienced severe food shortages. Food riots broke out in England and France, and grain warehouses were looted. Huge storms and abnormal rainfall flooded Europe's major rivers. A major typhus epidemic hit Ireland between 1816 and 1819, precipitated by the famine of the Year Without a Summer. An estimated 100,000 Irish perished during this period. In Europe, the fatality rates in 1816 were approximately 200,000, twice the normal average.

Cooler temperatures and heavy rains caused crop failures in Britain and Ireland. Families in Wales journeyed long distances begging for food. Famine was widespread in Ireland after farmers lost wheat, oats, and potato crops. The crises was also severe in Germany and food prices rose sharply. People rallied in front of grain markets and bakeries. Riots, arson, and looting occurred all over Europe. It was the worst famine of the 19th-century.

In Switzerland, the summers of 1816 and 1817 were so cool, an ice dam formed below the Giétro Glacier. Despite attempts to drain the growing lake, the ice dam crumpled tragically in June 1818. Hungary experienced brown snow and Italy saw red snow falling throughout the year caused by the volcanic ash in the atmosphere.

In northern China, the cold weather killed trees, rice crops, and water buffalo. Floods destroyed many of the remaining crops. Mount Tambora's eruption disrupted China's monsoon season, resulting in overwhelming floods in the Yangtze Valley. In India, the delayed summer monsoon caused late torrential rains that aggravated the spread of cholera from the Ganges River in Bengal to as far as Moscow, Russia. Summer snowfall and mixed precipitation were reported in various locations in Jiangxi and Anhui. In Taiwan, which has a tropical climate, snow was reported in Hsinchu and Miaoli and frost in Changhua.

In North America, during the spring and summer of 1816, an unrelenting dry fog covered parts of the eastern United States. It has been characterized as a "stratospheric sulfate aerosol veil." The fog reddened and dimmed the sunlight making sunspots visible to the naked eye. Even wind and rain could not disperse this "fog."

In May 1816, frost killed most crops in the higher elevations of New England and New York. On June 4th, frost was reported as far south as Connecticut and New Jersey. On June 6th, snow fell in Albany, New York, and Dennysville, Maine. Temperatures went below freezing almost every day in May. The ground froze solid on June 9th. On June 12th, the Shakers had to replant crops destroyed by the cold. On July 7th, it was so cold that everything stopped growing. As late as August 23rd much of the upper northeastern United States experienced this unheard-of frost.

A Massachusetts historian summed up the disaster: "Severe frosts occurred every month; June 7th and 8th snow fell, and it was so cold that crops were cut down, even freezing the roots …. In the early Autumn when corn was in the silk it was so thoroughly frozen that it never ripened and was scarcely worth harvesting. Breadstuffs were scarce and prices high and the poorer class of people were often in straits for want of food."

Farmers south of New England did bring some crops to maturity, but corn and other grain prices rose dramatically. The price of oats went from 12¢ a bushel to 92¢ a bushel. In the summer of 1816, corn ripened so poorly that only a quarter of it could be harvested for food. The crop failures in New England, Canada, and parts of Europe also caused the price of wheat, grains, meat, vegetables, butter, milk, and flour to rise sharply.

In June 1816, "incessant rainfall" during that "wet, ungenial summer" forced Mary Shelley, John William Polidori, and their friends to stay indoors for much of their Swiss holiday. They decided to have a contest to see who could write the scariest story. Mary Shelley wrote *Frankenstein* and Lord Byron wrote *A Fragment*, which inspired Polidori's *The Vampyre*, a forerunner to *Dracula*. Lord Byron also wrote his poem, *Darkness*, during this time.

The crop failures of the Year Without a Summer are also believed to have shaped the settling of the American Heartland, as many thousands of people (particularly farm families who were wiped out by the devastation) left New England for the Midwest in search of a more hospitable climate, richer soil, and better growing conditions. Among those who left Vermont was Joseph Smith, Sr.

The Smiths, desperate to make some headway in their battle with poverty and tenantry, left in the fall along with thousands of other Vermonters, and headed for Palmyra, New York. Palmyra was a fast-growing little village, soon to be connected with Albany and Rochester by way of the Erie Canal. There was ample work helping established farmers, and the Smiths did a small business in various trade items.

By 1818, after fifteen years of tenantry, the Smiths had contracted for a hundred-acre farm just two miles south of Palmyra over the boundary into Manchester Township. It was this move from Sharon, Vermont, to Palmyra, New York, brought on by the cataclysmic events of the severe climate change during the Year Without a Summer that precipitated the series of events that brought a humble farm boy, named Joseph Smith, Jr., to Palmyra, culminating in the publication of the Book of Mormon and the founding of the Church of Jesus Christ of Latter-day Saints.

I imagine that the religionists will reject the scientific claims of this work and that the scientists will reject its religious claims. We should remember that: "In the long search for truth, we have often reached wrong conclusions both in science and religion…. Life demands that we keep learning. Otherwise we become shunted off into dark pockets of error." [53]

---

[53] Franklin S. Harris, *Science and Your Faith in God*, Bookcraft, 1958, p 93.

I hope you enjoyed this little book. I would like very much if you could post an honest 5-star review on Amazon or some other book site where you have an account and posting privileges. Maybe you can mention what you liked best about it.

If you found this book enjoyable, inspirational, educational or enlightening, I would hope that you tell your friends about it.

About the Author

Bill Wylson is the author of over 50 published writings dealing with family values, religious issues and religious education. His work has appeared in *The Ensign, This People, Liberty Magazine, Success,* and others.

Bill graduated from the *Columbia School of Broadcasting* in Hollywood, CA as a commercial copywriter. He wrote trade journal ads for a major advertising agency in Los Angeles and public service announcements for a Los Angeles television station.

He has served as a volunteer Board Member of *Advocates of Single Parent Youth, Special Fun Games for the Disabled,* and on the Boards of Arts and Theater Councils. He has also served on Advisory Committees for the *Volunteer Center of Los Angeles* and on the *United Way Government Affairs Committee.*

Bill Wylson lives in Salt Lake City, Utah.

☐

Other Books by Bill Wylson

Three Minutes Eighteen Seconds:
A Prophet's Final Message to the World

Words are extremely powerful. Lord Byron poetically portrays this truth:

"But words are things, and a small drop of ink,

"Falling like dew, upon a thought, produces

"That which makes thousands, perhaps millions, think."

Three Minutes Eighteen Seconds examines three "small drops of ink" that are, simultaneously, extremely powerful words spoken by President Thomas S. Monson in the April 2017 General Conference.

Hieroglyphs, Golden Plates and Typos:

On the inside cover of his first leather-bound Book of Mormon my father had written the following quotation from the prophet Joseph Smith: "I told the brethren that the Book of Mormon was the most correct of any book on earth, and the keystone of our religion, and a man would get nearer to God by abiding by its precepts, than by any other book." Directly below this quote, my father had compiled a list of scriptures which he had labeled: "Mistakes in the Book of Mormon."

Committing his writings to the future reader, Moroni candidly and apologetically acknowledged: "And if there be faults they be the faults of a man. But behold, we know no fault."

How then did my father have the boldness to make a list of mistakes in the Book of Mormon? To gain a better understanding of these 'corrections' in the Book of Mormon and how they testify to its truthfulness and authenticity, we need to understand the process involved in making plates of ore and the method for inscribing on them.

Elder Hammond and the Inspector

"You know, there's a word to describe someone who won't even bother to meet his new companion at the bus station. It starts with an 'O' or, I don't know, maybe a 'C' or something. I think it's C-a—. No, I've lost it."

Elder Hammond was a freckled-face, shy sort of bumpkin from some rural farm town in Kansas. He was awkward and withdrawn. Even in his white shirt and tie he reminded you of the type of kid you'd see in denim coveralls, wearin' a straw hat and chompin' on a thin blade of grass whilst irrigatin' the lower forty.

I actually knew nothing about Elder Hammond's personal life. He was just a simple, quiet, humble boy but he was also determined and dedicated. He had no delusions of grandeur, just a desire to serve. Perhaps more than any missionary I had ever met, Elder Hammond had a purity of spirit and an altruistic motivation in ministering. I pitied him. I think he actually believed he could make a difference.

The Manger on the Mantle
A Christmas Tale based on Two True Stories

The Manger on the Mantle recounts the tragic life of Mark Spencer, a man raised with an idyllic small-town childhood who somehow becomes very lost in the massive city of Los Angeles. The problem is he didn't become geographically lost; he became spiritually lost.

As his family falls apart and his world collapses, Mark begins to realize just how tainted and disgraceful his life has become. He has strayed so far from the innocence of his youth and now he fears he may never find his way back.

That's when Mark meets a sockless, root-beer-float toting ex-hippie by the name of Marvin. Together they journey the road to Bethlehem as they ponder the purpose of a birth that took place in a lowly manger.

The Manger on the Mantle is a beautiful story of hope and redemption and the joyous possibility of being given a second chance.

Give Place in Your Heart:
31 Promises from the Book of Mormon

All of us are familiar with Moroni's promise that Christ will manifest the truth of the Book of Mormon to us by the power of the Holy Ghost. This is just one of many promises the Lord has made regarding the Book of Mormon.

In *Give Place in Your Heart*, Bill Wylson outlines 31 promises, with their attendant blessings and conditions, that the Lord would love to bestow upon you.